JOHN DEERE TRACTORS

Jonathan Whitlam

AMBERLEY

First published 2018

Amberley Publishing
The Hill, Stroud,
Gloucestershire, GL5 4EP

www.amberley-books.com

ISBN: 978 1 4456 6784 3 (print)
ISBN: 978 1 4456 6785 0 (ebook)

British Library Cataloguing in Publication Data.
A catalogue record for this book is available from the British Library.

Typeset in 10pt on 13pt Celeste.
Origination by Amberley Publishing.
Printed in the UK.

Contents

Preface

This book is not intended as a full technical history of the John Deere tractor or the John Deere Company but is a broad overview showing the evolution of the tractors that have held the John Deere name for nearly a century.

This book is not an official publication and is not endorsed in any way by John Deere. We recognise that the John Deere name and various other brand names are the copyright of Deere & Co.

All horsepower outputs and similar figures are approximate and should only be used as a rough guide. They are taken from contemporary literature of the time. All views expressed are my own, as are any inevitable errors that might have crept into the text.

Introduction

John Deere is a unique company. Despite all the upheavals of modern times, this tractor, farm and horticultural machinery and construction equipment builder has managed to stay totally independent all of its life, unlike its competition, which have all been the subject of many mergers and takeovers, particularly during the 1980s and '90s.

John Deere has remained a constant throughout this, and although tractors are only a small part of its business, the story of the green and yellow machines is a fascinating one that covers the entire history of the farm tractor itself.

The badge used on the equipment has changed over the years, but the John Deere name hasn't.

John Deere was the man who started it all, hence his name still being used to this day. He started a blacksmith's shop in Grand Detour, Illinois, back in 1836, making his very first steel plough the following year. By 1859, which is when he moved to Moline, also in Illinois, a wide range of cultivation and planting equipment had been added to the range of implements he was building. Although John Deere died in 1886, the business he started would live on.

It was in 1892 that the John Deere company first began to tinker with the idea of producing a tractor, including building a number of unusual looking three-wheel machines that were known as the Dain All Wheel Drive Tractor, named after a member of the board, Joseph Dain, who developed the machine. Despite actually working, and 100 of these tractors being built between 1918 and 1919, they sadly proved to be far too expensive to build and tractor production stopped there.

However, in 1918 the company made its first real breakthrough in the tractor business and acquired an outside firm that was already building a successful tractor. This would be the start of one of the most successful tractor stories of all time.

CHAPTER 1

Waterloo Boy Sets a Trend

The Waterloo Gasoline Tractor Co., based as it name suggests in Waterloo, Iowa, were making the Waterloo Boy Model N and Model R tractors. These were simple machines but also very reliable ones for their time, being powered by two-cylinder engines mounted on a large frame that also supported the transmission and radiator. This was very much the traditional way of building farm tractors, following on largely from steam traction engine practice, but it worked.

The Waterloo Boy was the tractor that got John Deere into the tractor business – and it also gave the company its livery and the two-cylinder engine.

The early versions of the Waterloo Boy, launched as early as 1914, were twelve drawbar horsepower tractors while the later Model R provided sixteen drawbar horsepower and 25 hp at the belt from its two-cylinder engine driving through a single-speed gearbox. The Model N replaced the earlier tractor in 1917 and now included a two-speed gearbox.

The Waterloo Boy was even sold into Britain, where it was called the Overtime and painted in the rather less attractive colours of green, grey and orange rather than the green and yellow of the native machine. It was sent over in bits and was assembled in the UK, with a total of 4,000 being built.

Firms such as Wallis and then Henry Ford had shown the way forward for tractor design even before John Deere bought the Waterloo Gasoline Tractor Co. in 1918, with the various Wallis tractors using a type of hull under the engine running to the back of the tractor, into which all the transmission components were positioned. This not only protected them from dust and dirt, but also provided the 'backbone' of the tractor, doing away with the need for a heavy subframe. The Fordson Model F, first produced in 1917, also adopted this method of unit construction, and with the mass-production facilities of Henry Ford behind it soon came to dominate the American market.

The Waterloo Boy was undoubtedly a huge success for John Deere; not only did it get the firm into the tractor business, it also gave them their corporate colours of green and yellow!

John Deere needed to modernise their inherited tractor design and this they did, but they kept one very important ingredient – the two-cylinder engine.

The Waterloo Gasoline Engine Company of Waterloo in Iowa had this beautifully designed logo on the front of their tractors.

Both these tractors, seen at a rally in Lincolnshire, were built by the Waterloo Boy company, but the one on the left is the version sold into the UK at the time and was called the Overtime.

This 1919 vintage Overtime is seen at a rally in Suffolk complete with a riding plough behind it.

The basic operator's platform of the Overtime showing the steering wheel and pan type seat.

The front of the Overtime tractor bore a very different logo to that of the North American Waterloo Boy and was designed to show the British farmers that it could be relied on to keep on working.

CHAPTER 2

Two-Cylinder Power

In 1923 the John Deere tractor proper appeared – the Model D. This was the first ever tractor to carry the now famous name, although the design was already on the drawing board at the Waterloo company when John Deere acquired the business.

Fitted with a horizontally mounted two-cylinder engine, the Model D set a trend for the next four decades of green and yellow tractors and also made use of the unit construction principle to give a lighter, more user-friendly machine that sat lower to the ground and also meant a shorter wheelbase. With 27 hp at the belt and 22½ hp at the drawbar, the Model D was considerably more powerful than the Waterloo Boy.

The Model D was improved in 1935 with more power and a three-speed gearbox, with further improvements in 1939 making it an even more powerful tractor. Built for a total of thirty years until 1953, the Model D is the world's longest running production tractor, by which figure its true success can be measured.

In 1927, John Deere decided to take on the popular International Harvester Farmall row crop tractor with the lighter 10 drawbar horsepower and 20 belt hp Model C. This tractor was designed to work in crops and was fitted with a three-speed gearbox and even a mechanical implement lift.

However, the Model C was not a success, and teething troubles led to many having to be recalled to the factory for rectification. This in turn led to a better effort in the shape of the John Deere General Purpose, or GP, that followed immediately after the Model C.

In 1934 the Model A appeared; a tricycle-type tractor design, which was available in row crop, narrow, wide, orchard and high-clearance versions. A mechanical lift mechanism for implements was included, as were adjustable track rear wheels. A four-speed transmission was now used but the two-cylinder engine was still very familiar.

The Model B, introduced in 1935, was a smaller version of the A and was rated at 16 hp. Both tractors were designed to replace the GP, which they did very successfully.

It was also in 1934 that the AR and BR were introduced as general-purpose variants of the Model A and B and were offered, with a choice of either steel wheels or pneumatic tyres.

The Model D was the first John Deere tractor proper, but had been on the Waterloo design boards when John Deere took over the company. Note the old tyre treads fitted to the front steel wheels of this preserved example.

Unitary construction was used to build the Model D, unlike the frame used for the Waterloo Boy, but the two-cylinder engine remained.

The John Deere Model D was not only the start of a tractor dynasty that survives to this day, so too do many of the tens of thousands of Model Ds that were built.

As demonstrated by its wide wheel track and narrow steel wheels, the General Purpose was designed to work among the growing crops.

This GPWT has V-twin front wheels and is seen on display at the Great Dorset Steam Fair.

Various different versions of the A and B replaced the General Purpose. This standard track AR is ploughing in the south of England. (Photo: Kim Parks)

Lindemann built crawler conversions of John Deere tractors right from the Model D and was eventually taken over by the company.

Displayed at a steam rally in Suffolk, this General Purpose has been restored to a very high standard.

The AOS was a special low version of the A and came complete with metal shields for working in orchards.

In 1937 the Model L came along; a small tractor designed as a replacement for a single horse and powered by a vertically mounted Hercules engine, but still of the two-cylinder type. Fitted with a three-speed transmission, the L also featured an offset seat and steering wheel to allow better visibility when working in row crops. The Hercules engine was replaced by a 10 hp John Deere unit in 1942.

A new four-speed transmission was introduced in 1938 to the Model A and B tractors but this was eclipsed somewhat by the new styling also introduced at the same time, and which modernised the look of the John Deere tractor line immensely. John Deere had turned to industrial designer Henry Dreyfuss to produce a new family look for their tractors. Instead of just using the bare mechanicals to fabricate the design of the tractor, in future they would all be finished with properly designed tinwork. The end result was not only good to look at, they were also extremely practical for everyday farm work and maintenance. From now on the new-look tractors were known as 'styled' and the earlier machines were known as 'unstyled'.

This new look would be rolled out over the entire tractor range but the first new models to carry it were the 36 hp Model G of 1938 and the 14 hp Model H of 1939. Both were row crop tractors and were fitted with a four-forward and single-reverse transmission. The Model G was the largest row crop John Deere tractor so far. Later versions were equipped with a high and low transmission giving six ratios.

The LA, sold alongside the L from 1940, featured a 12 hp engine and a solid bar frame and also had the option of a rear PTO, adjustable front axle and electric starting. As the smallest tractor in the range, the L and then the LA represented only a niche position in the tractor market as a whole.

The smaller end of the tractor market was not neglected by John Deere, who produced the Model L and LA to cater for market gardeners and smaller acreage farmers who did not need the power of the larger machines. This preserved example is shown with a mid-mounted finger bar mower.

The Model A was one of the first tractors to benefit from the new styling brought in by Henry Dreyfuss. It certainly made the tractors look more modern, as well as larger.

The Model B was basically a slightly smaller version of the Model A. Note the row crop configuration with a single front wheel, as well as the long steering rod across the top of the bonnet.

A wide-track Model B with two front wheels showing just how wide the axles could be set for clearing row crops.

The Model G was a larger tractor and as such was the biggest row crop model built by John Deere up to this point.

A restored Model H showing the V-twin front wheels often preferred by many farmers in the United States.

In 1947 the Model M arrived with a 20 hp horizontal two-cylinder petrol engine and also such modern refinements as electric starting, power take-off, lights and the new Touch-O-Matic hydraulic system complete with a three-point linkage to attach to suitable implements. The Model M actually replaced as many as five different previous tractor models and was also used as the base for the MT tricycle wheel row crop, MC crawler and MI industrial machines.

The MC crawler version was actually built by a firm called Lindemann, which had been taken over by Deere in 1945 after making crawler versions of many John Deere tractors as far back as the Model D.

The Model R, introduced in 1949, was a landmark machine for the company. This was the first John Deere diesel tractor. Its 48 hp two-cylinder power plant was started by a petrol donkey engine and, amazingly, it was also the first American tractor to be fitted with a full diesel engine, being equipped with a five-speed transmission and the new and improved Power-Trol hydraulics. The Model R looked an impressive machine as befitted its credentials, and this was emphasised by the very large front radiator grille with vertical bars.

The Model R would be the last of the letter series that began with the Model D back in 1923. From 1952 new models would be identified by numbers, starting with the 31 hp 50 and 41 hp 60 models. Two-cylinder engines would remain the norm, however, even though competitive manufacturers had turned to four- and six-cylinders, even back in the days of the Model D. John Deere stuck to its roots though, and two-cylinders remained the preferred format throughout the new models. Behind the scenes it was a different story and work was underway on a whole new range of multi-cylinder tractors for the future.

The Model M was John Deere's answer to an all-round versatile machine and was introduced in 1947. The M was the base model for a range of various different types.

V-twin front wheels are fitted to this tricycle version of the M, the MT, which is shown on display at a rally in Norfolk.

Even a crawler version of the M was produced, called the MC and owing much to the earlier Lindemann tracklayers.

The Model D soldiered on for several decades right into the 1950s and was updated several times, including with the new Dreyfuss styling and more power.

A Model R at work in East Sussex with a cultivator. This was the first diesel tractor from John Deere and was the most powerful so far. (Photo: Kim Parks)

Above: Late Model R tractors adopted a revised livery with yellow on the sides of the bonnet.

Left: Numbers replaced letters in the early 1950s and this 60 is a very high-clearance machine.

Back to the early 1950s and the 50 and 60 were joined by the John Deere 70 and 40 in 1953, the new and improved 70 in 1954 and the 80 in 1954 (the latter replacing the Model R and fitted, as were the other models, with a six-forward and single-reverse gearbox). New on the number series was a live power take-off and a new type of rear axle design. Otherwise, these tractors were very similar to the letter series models they replaced, the new styling making them look larger than they actually were. Liquid petroleum gas was also able to be used as a fuel thanks to LPG versions, with this being a popular choice for many farmers as the '50s went on – it being a cheaper fuel than petrol but not as efficient as diesel.

There were also two diesel-powered choices in the range now. These were the 51½ hp 70 and 69 hp 80, both of which were available with this engine alternative, such was the success of the 80's predecessor, the Model R.

The period between 1955 and 1956 saw further model changes when the 40 became the 320 and was joined by the new 29 hp 420. The rest of the range joined these; the 520, 620, 720 and 820, featuring draft control hydraulic systems and a more comfortable driver's seat as standard, but the two-cylinder engines remained.

The restyled John Deere 30 Series followed in 1959 and included the unusual 435, fitted with a General Motors Detroit two-cylinder diesel engine of 33 hp. Needing a smaller diesel tractor in the range but with the New Generation tractors on the horizon, John Deere decided to go down the route of buying in an 'off the peg' diesel engine instead of developing their own unit for what would be a tractor with a short production life due to the new developments of 1960.

This 70 is fitted with a very rudimentary looking John Deere front loader and looks to be in its original condition.

The mid-1950s saw the likes of the 320 arrive, which replaced the 40. Note the increased use of yellow on the tinwork and the rear wheel weights on this example.

Liquid petroleum gas became a very popular choice for powering farm machinery in the USA during the 1950s, as demonstrated by this restored 620 model. Also note the large-diameter chrome exhaust stack.

The 820 was the descendent of the Model R and, as can be seen here, did not change much physically from the earlier tractor.

1959 saw the 30 Series arrive including the 830 – a tractor that was seen by many as a very large machine back then.

The 435 was unusual in that it was not powered by a John Deere two-cylinder engine; instead, a General Motors two-cylinder was used.

Another view of an 830 Diesel showing the large front radiator grille.

CHAPTER 3

New Generation

John Deere made a big step in its worldwide sales when it acquired a major share in the Lanz concern based in Germany in 1956. Famous for its line of single-cylinder Bulldog tractors built in Mannheim, the John Deere influence saw the first multi-cylinder tractors appear, which were known as the John Deere-Lanz 300 and 500 in 1961.

However, these were not the first John Deere tractors to use more than two cylinders, as back in North America the New Generation 10 Series was launched in 1960. Both wheeled and crawler versions of the 35 hp 1010 and 45 hp 2010 were offered, powered by four-cylinder Deere engines with the choice of petrol or diesel fuel, and the 2010 also having an LPG version. These two tractors were built in Deere's Dubuque factory in Iowa, which had originally opened to build the Model M.

John Deere had been firmly wedded to the two-cylinder engine and Lanz in Germany had been similarly dedicated to single-cylinder power with their Bulldog line.

The takeover by John Deere would change all of that, with the multi-cylinder 300 and 500 being the first. This John Deere-Lanz 500 is shown in Lincolnshire.

In the USA, the New Generation tractors saw the two-cylinder engine consigned to the scrapheap. The 1010 was one of the first, launched in 1960, and this is a rare industrial version of it.

The rest of the range was built in Waterloo in Iowa with the four-cylinder 55 hp 3010 and six-cylinder 80 hp 4010 arriving in 1961, followed by the six-cylinder 5010 with 121 hp in 1962.

With new multi-cylinder engines (up to six in the largest two tractors), the new range swept away the traditional two-cylinder engine used since the beginning of John Deere tractor production. It was a radical move but one that was needed as the 'Johnny Popper' two-cylinder tractors were looking rather dated next to the competition, all of whom had begun using larger multi-cylinder engines long before 1960. The New Generation, as it was called, really did look as new as they actually were, with a brand-new style including a curved bonnet, distinctive frontal treatment and a raft of new features.

The eight speed Synchro-Range transmission was one of the new features and gave a wide spread of ratios controlled by levers built into the steering wheel binnacle. The 4010 was the first John Deere tractor with the company's own six-cylinder engine and remained the only one until the 5010 arrived, which was the largest two-wheel drive production tractor at the time. John Deere not only replaced all their existing tractor models, they did so by producing tractors that were bigger and better than the competition, and all were introduced to the market in just two years.

Barely had the new tractor design settled down than the next generation was being introduced, with the 3020 and 4020 appearing in 1963 and being followed by the 1020 and 2020 in 1965 and the 1120 and 5020 in 1966.

The 4020 was the tractor that really saw John Deere become a force in the UK and it sold well in many other countries around the world as well.

The 4020 tractor in particular proved a very popular machine, especially in the UK where it had no real competition back in 1964, as no other native machine came anywhere near its power and level of sophistication. Fitted with a six-cylinder 100 hp Deere diesel engine and an eight-forward and two-reverse synchromesh transmission, as well as power steering and a two-speed power take-off, this was a real revelation to many British farmers – or at least those with enough acres to justify it! The 4020 was further enhanced when an optional eight-forward and four-reverse full powershift transmission became available, making this a more modern tractor than its production date would suggest, with the gearbox allowing on-the-move shifting through all ratios. The four-cylinder 82 hp 3020 could also be bought with the powershift transmission as an alternative to its standard synchromesh box.

Meanwhile, tractors were still being produced in Mannheim in Germany, including the John Deere 710 with 56 hp on offer and with similar styling to the American machines, and this was followed by the German factory building the three-cylinder 47 hp 1020, 53 hp 1120 and 64 hp 2020 tractors – the latter having four-cylinder Deere engines. Closed-centre hydraulic systems, differential lock and disc brakes were among the list of standard features on these tractors, along with optional extras such as dual speed power take-off, power steering and a Hi-Lo change on the move gear splitter. 1969 saw the first six-cylinder tractor to be built at the Mannheim factory with the introduction of the 3120 model. Production also continued in the USA and the range was joined by the 2520 and 4520 in 1969.

The dashboard of the 4020 featured clear instrumentation and the controls for the transmission – in this case the groundbreaking eight-speed powershift.

A very comfortable, fully padded seat was fitted to the 4020, giving excellent support for the operator.

A smooth six-cylinder Deere engine of 100 hp provided the grunt for the 4020 – a type of power that was unheard of on many British farms at that time.

Four-wheel drive versions of the 4020 are not common, and this one seen at the Cheffins auction in Cambridgeshire probably hails from Europe.

The little 1020 was built in the USA as well as in Germany at the Lanz factory in Mannheim.

The distinctive styling used on the 1020 was first seen on the 10 Series tractors launched in 1960 and was carried on through the 20 and 30 Series that followed.

The 2020 first arrived in 1965 and was a popular machine with its four-cylinder Deere engine.

This 1120 dates from 1972 and is still used for ploughing matches.

The 5020 was the conventional flagship of the 20 Series, this six-cylinder tractor being an extremely large and powerful one. Although several were sold in the UK, not many farmers could justify this amount of grunt.

The 3120 was a landmark machine as it was the first six-cylinder tractor to be built at Mannheim. It still used the styling that had been used for the rest of the range. (Photo: Kim Parks)

The 2120 was later added to the range of Mannheim-built machines.

The 6030 was a very large tractor indeed and the biggest rigid-frame machine so far built by John Deere. Designed with the large open prairies of the USA in mind, this tractor was a much later import into the UK.

Generation II tractors arrived during 1972 with the high-horsepower six-cylinder 30 Series models built at Waterloo. New bonnet styling was teamed with a new standard Quad-Range sixteen-speed transmission, although the powershift gearbox was still an option. Starting at the 80 hp 4030, the range also included the 100 hp 4230, 125 hp 4430 and the 150 hp 4630, all being powered by the same six-cylinder engine but with the two largest models featuring a turbocharger. The main feature was the new Sound-Gard quiet cab – a sound insulated cabin that gave the best noise protection in the business and was way ahead of the competition, offering sound deadening plus roll-over protection.

In 1975 Mannheim was producing smaller 30 Series tractors, most often being fitted with Duncan cabs to meet the safety cab legislation in Britain introduced in 1970. However, to meet the quiet cab rules of 1976 a Sekura-built OPU cab was fitted. OPU stood for Operator Protection Unit. The German-built 30 Series also now had the option of hydraulic front-wheel assist on the 2130 and 3130 models, while from 1976 mechanical front-wheel drive was also an option on the smaller models.

1972 saw the Generation II tractors arrive, with the 125 hp 4430 being the second largest in an all-new range that also included the market-leading Sound-Gard cab.

This Irish example of the 4430 is also fitted with the hydraulically operated four-wheel drive assist front axle.

Mannheim produced smaller 30 Series tractors such as this 1130, which also featured revised styling.

This 1630 with narrow rear row crop wheels is also fitted with a Duncan safety cab, as was used on many 30 Series tractors in the early 1970s.

The John Deere OPU cab, fitted to this 2130, was built for the company by Sekura and was built to pass the noise level legislation imposed on the tractor industry in the UK in 1976.

Many John Deere 30 Series tractors are still earning their living on farms, as is this 3130, seen rolling in East Sussex, which is also fitted with the OPU cab. (Photo: Kim Parks)

CHAPTER 4

Giants in Green

John Deere was pretty early to recognise the need for high-horsepower four-wheel drive tractors, thus producing the six-cylinder 8010, which was an articulated steer four-wheel drive monster. The Detroit Diesel engine produced 215 hp – an unheard of figure in 1959 when the 8010 was launched – and that enormous power was put to the ground through a nine-speed transmission powering all four wheels. Teething troubles with the design led to many being recalled and then modified as the 8020, now incorporating a strengthened eight-speed transmission that was more up to the job than the original unit fitted. Unfortunately the tractor did not sell well, probably because of how expensive it was, and even in 1965 some 8020 tractors remained in stock having failed to find a buyer.

Following this rather inauspicious start, John Deere entered into an agreement with pioneer articulated four-wheel drive tractor maker Wagner to produce its tractors exclusively in John Deere livery and with the John Deere name on the bonnet. This saw the Wagner 225 hp WA-14 and 280 hp WA-17 being produced, painted in John Deere colours with John Deere badging and sold through the dealer network from 1968, but this was only ever intended as a stopgap measure.

Indeed, the Wagner deal did not last long before new articulated tractors appeared in the shape of the 146 hp 7020 and 175 hp 7520, built in-house by John Deere and proving much more successful than the earlier attempts. The 7020 arrived first in 1970 with its Deere six-cylinder, turbocharged and intercooled engine, which was followed in 1972 by the similar but more powerful 7520. These two tractors were the start of what was to become a true success story for John Deere in the world of articulated four-wheel drive tractors.

These were followed in 1975 by the much-improved 215 hp 8430 and 275 hp 8630 monsters – bigger machines that were also fitted with the Sound-Gard cab. The gearbox fitted in these large machines was the Quad-Range sixteen-forward and four-reverse transmission, while the six-cylinder Deere engines featured both turbocharging and intercooling.

These evolved into 40 Series versions in 1979 with several improvements being made to the cab layout and other items, including a single lever to control the Quad-Range transmission and also hydraulically controlled differential locks.

When John Deere finally came up with their own successful articulated pivot-steer tractors, the largest of the lot was the 7520. This example is shown in its native USA complete with dual wheels all round and a front blade. (Photo: Paul Reeve)

The 8430 saw a much better package than the earlier 20 Series, with more weight and power and also the Sound-Gard cab. (Photo: Kim Parks)

An 8440 in the snow in Suffolk with its dual wheels removed.

The 8640 is still an impressive machine and this one is pulling an eight-furrow plough on hilly land in East Sussex.

Right: The simple control lever for the Quad-Range transmission in the cab of the 8640.

Below: The design of the articulated 40 Series tractors was based heavily on the earlier two 30 Series models and was ideal for getting its large power down to the ground, especially when fitted with dual wheels all round.

CHAPTER 5

The Task Masters

From 1977 the first 40 Series tractors appeared, called the 'Iron Horses' with four high-horsepower tractors spanning from the 100 hp 4040 to the 177 hp 4640. Built in the USA these tractors featured the new SG2 cab – an upgraded version of the original Sound-Gard unit. All were six-cylinder tractors and could be had with optional four-wheel drive (the original hydraulically operated front wheel assist later being replaced by a much more reliable and efficient mechanical four-wheel drive system).

The 40 Series saw the improved SG2 cab but otherwise the styling remained much as before. Two-wheel drive versions of the 4040 were quite common and were certainly the most popular in North America. This fabulous example is shown in Ireland.

Mannheim also produced 40 Series tractors, the 3040 being a very popular model and also fitted with the SG2 cab. (Photo: Kim Parks)

From 1979 Mannheim was also producing 40 Series models, called the 'Schedule Masters' by the company, and spanning no less than seven models from the little 50 hp 1040 to the 97 hp 3140. These tractors came with an eight-speed synchromesh gearbox as standard although the Power-Synchron high and low gear splitter was available at extra cost. Four-wheel drive was offered as an option on the larger models in the range and the SG2 cab was now also an option.

In 1981 the 115 hp 4040 and 132 hp 4240 with turbocharged six-cylinder motors were available, having been assembled in Mannheim using mainly North American parts, and could be fitted with a ZF-sourced front-drive axle. This was particularly popular in Europe although back in their native USA they were more often sold as two-wheel drive machines.

The early 1980s saw the 40 Series expand quite quickly with new, smaller, low-profile tractors and an XE economy version of many models in both the German and USA model ranges.

More high-horsepower machines arrived in 1982 when the first of the 50 Series made their appearance, with smaller and medium-sized tractors later taking this range to sixteen models from 44 hp to 370 hp. The turning angle of the Waterloo-built 100 hp 4050 to 190 hp 4850 tractors was improved by the fitting of a new front-drive axle with a castor action.

The 370 hp 8850 articulated monster was the largest John Deere tractor so far and was powered by a V8 engine. From the 140 hp 4350 up to the giant 8850, a new touch-operated digital performance monitor was fitted into the cab, bringing the John Deere tractors into the electronic age for the first time.

The 4240S was an improved version of the Iowa-built 4040 and this one is using a New Holland 525 forage harvester in Derbyshire.

The SG2 cab gave excellent all-round visibility, including out of the large rear window.

Above: This 2140 with four-wheel drive is fitted with the OPU cab option and is taking part in a ploughing match in the south of England. (Photo: Kim Parks)

Right: Many 2140 tractors were fitted with the SG2 cab option and this one shows off the tight turning circle made possible by the design of the front axle.

Above: Another 2140, this time of two-wheel drive layout, seen ploughing in Suffolk.

Left: The XE range was an economy version of the Mannheim-built tractors. This 2140 XE is awaiting a new owner at auction.

The 3640 was the largest tractor built at Mannheim in the 40 Series era. Breaking up ploughed land in Suffolk with a Wilder pressure harrow is well within the capabilities of this large six-cylinder tractor.

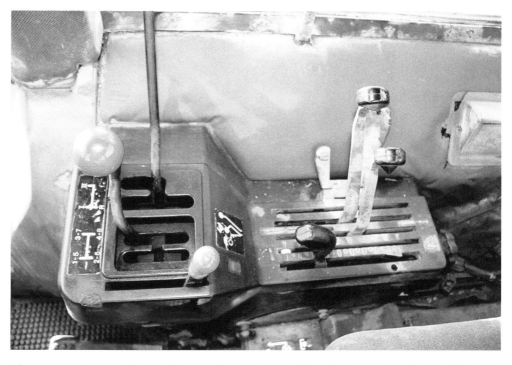

The gear quadrant and hydraulic controls to the right of the driving seat in the 3640 SG2 cab.

The 4850 was the largest rigid-frame John Deere tractor for a time with 190 hp on tap. Dual rear wheels and a front linkage are fitted to this example in East Sussex.

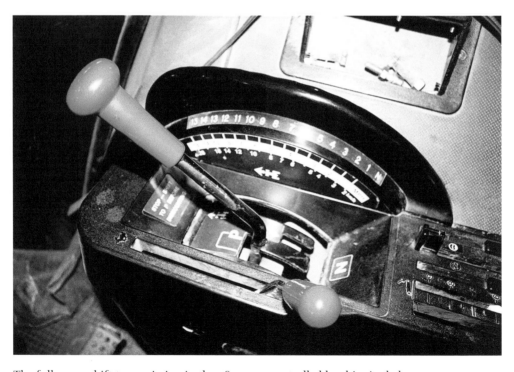

The full powershift transmission in the 4850 was controlled by this single lever.

The smaller end of the 50 Series were sourced from Mannheim and began to come on stream in 1986 with three-, four- and six-cylinder tractors spanning from the 50hp 1750 up to the 100 hp 3350. A new 40 km/h version of the Power Synchron sixteen-speed transmission was standard on the six-cylinder models and was available as an option on the four-cylinder tractors. The bigger 3650 arrived later.

Fifteen-speed powershift transmissions were the order of the day on the new American-built 128 hp 4050 and 144 hp 4250 tractors, which were also sold in the UK from 1987. The full powershift gave completely clutch-free operation of all the gears in either forward or reverse, and was truly a world-leading system.

More big tractors arrived in 1990 when the 128 hp to 228 hp 55 Series arrived, making the high-horsepower ratings of articulated tractors over 200 hp available in a rigid design for the first time. Full powershift fifteen-speed transmission and the John Deere Intellitrak Monitoring System made these very sophisticated tractors, monitoring the performance of the tractor in the field and displaying the information garnered to the operator as well as also being used by service technicians to make diagnostic checks on the machine.

60 Series models made an appearance in 1989 with the 235 hp 8560 and 300 hp 8760 having six-cylinder Deere engines while the largest, the 370 hp 8960, used a 14 litre Cummins six-cylinder motor. These behemoths came complete with electronic engine governing and remained in place until the four-model 70 Series replaced them in 1993, with the range topping at the giant 400 hp 8970.

When it came to the main-line smaller machines, however, it was the 50 Series that would truly be the final flowering of a design that could trace its history back to 1960 and even further, right back to 1923. 1992 would see the beginning of the end of that legacy.

The 8650 was a large articulated powerhouse and is shown here fitted with large flotation tyres for top work. (Photo: Kim Parks)

Bigger articulated John Deere tractors were popular in the USA, including this 8760 model, shown at a large demonstration event. (Photo: Paul Reeve)

The little 1850 made in Germany was ideal for row crop work when fitted with narrow wheels all round.

The 2650 was a very popular tractor in the UK.

The bigger 2850 was also common and this one is fitted with a Howard front loader and rear flotation tyres for work in Suffolk.

Inside the SG2 cab of the 2850, the comfortable and fully adjustable seat can be seen with all the main controls mounted up to the right-hand side.

A 2450 and John Deere loader baling in Suffolk.

A 3050 ploughing in East Sussex with a Dowdeswell plough. (Photo: Kim Parks)

A stunningly restored 3350 in two-wheel drive form at a show in Lincolnshire.

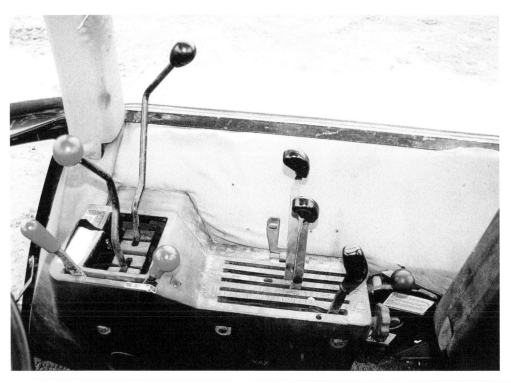

The control console to the right of the seat in the SG2 cab fitted to a 3350.

The 3650 was the largest Mannheim-built tractor in the late 1980s and early '90s and this one is shown competing in a ploughing match in southern England. (Photo: Kim Parks)

The American-built 55 Series of six-cylinder tractors were all large, and the 4055 was actually the smallest. (Photo: Kim Parks)

Looking impressive with dual rear wheels, this 4455 is still very much a working machine on a farm in Suffolk and is used primarily for cultivation and carting duties.

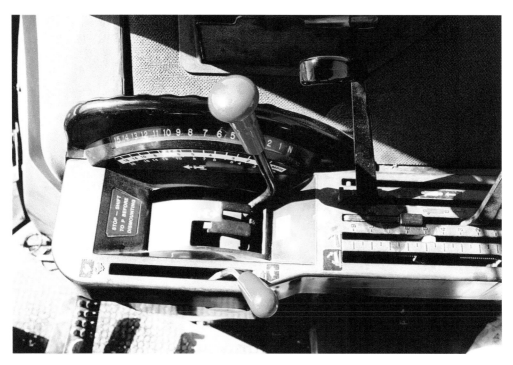

A single lever controlled the powershift transmission in the 4455, making it simple to operate.

The 4955 was the biggest of the lot and survived into the mid-1990s complete with an exhaust stack mounted up beside the front cab pillar. The large bonnet concealed a very powerful 228 hp six-cylinder Deere engine.

CHAPTER 6

Let's Reinvent the Tractor

In 1923 the Model D was the first John Deere tractor to do away with a supporting frame for the engine and other components, but in 1992 the company changed this policy and went back to what it called a 'full-frame' design.

This essentially meant that the new 6000 and 7000 Series tractors used a type of chassis frame to which everything was mounted instead of using the unitary construction method used universally by the industry for many decades. These two new tractor ranges were also milestones for other reasons; they were new from the ground up, with totally new cabs, engines and transmissions, to name but a few features.

The smaller of the two new ranges were built at Mannheim in Germany and consisted of four models with Deere four-cylinder engines spanning from the 6100 at 75 hp to the turbocharged 100 hp 6400 and featuring a choice of either the SynchroPlus or PowerQuad transmissions – the latter being a four-stage semi-powershift gearbox. The TechCenter cab was a completely new venture for John Deere as it replaced the SG2 cab, and for the first time did not include the by-now iconic split front windscreen with left-hand door. Now there were opening doors on either side of the cab, larger areas of glass and the cabin was moved further forward on the chassis for a smoother ride. The exhaust stack was also moved from the top of the bonnet to the right-hand side front cab pillar, thus greatly improving forward visibility.

All the controls were placed up to the right of the driver on a console and the instrument binnacle around the steering wheel was kept as small as possible to help with forward visibility. Full electronic monitoring and electronic control of the hydraulic systems was also included.

The bigger three 7000 Series models had similar features and were built in the USA. These ranged from the 130 hp 7600 up to the 170 hp 7800, all coming with six-cylinder Deere engines. This was basically the situation in Europe, but as always different countries were offered different versions, which even saw the 7000 range being brought down to meet the 6000 Series with the likes of the 7200 and 7400 machines. An electronically, single-lever-controlled, full-powershift, nineteen-forward and seven-reverse transmission was available as an option on these big tractors that replaced all but the largest 55 Series models.

Smallest of the new ranges launched in 1992 was the 75 hp 6100, an example of which is picking up hay bales in East Sussex with a John Deere loader. (Photo: Kim Parks)

The 100 hp four-cylinder 6400 was the largest 6000 Series model when launched and was something of a pocket rocket with plenty of power in a compact package. A Standen Eho potato planter is behind this one, seen working in Kent. (Photo: Kim Parks)

The middle model in the new 7000 Series was the 7700. The full-frame design was ideal for fitting front linkages to the front of the frame. As can be seen, the styling was shared with the 6000 Series, but these American-built tractors were substantially larger.

The 170 hp 7800 was the largest tractor in this range. Note the adjustable bar-type rear axle to vary the track width – a typical American fitment.

These surviving 55 Series tractors were the 190 hp 4755 and 228 hp 4955 – really big machines with a rigid frame design but plenty of power. These were sold alongside the new 7000 Series and were unusual in that late models had their bonnet-mounted exhaust stack moved to the right-hand side front cab pillar, but the SG2 cab that they were fitted with still had the central pillar! These two big tractors soldiered on until 1995, when the 185 to 260 hp 8000 Series finally replaced them and once again used a full-frame design. The engine was now mounted right at the front of the chassis instead of nearer the cab, giving a better weight distribution between the front and rear axles, while all the engines were turbocharged and intercooled six-cylinder Deere units. Up in the spacious new cab the operator could experience previously unheard of levels of comfort and visibility thanks to a side-mounted digital display, which meant that the steering binnacle was now just that, a steering wheel, giving superb uninterrupted forward vision across the long green bonnet.

The seat was fitted with the CommandArm on the right-hand side armrest for all the main functions, including a sixteen-speed full-powershift transmission.

Further models joined the original 6000 Series tractors in 1993, with two six-cylinder models joining the existing 'four-pot' tractors. The 6600 was a 110 hp machine while the new 6800 had 120 hp. Apart from their larger engines, these two new models had exactly the same features as the earlier tractors. These were also later joined by another even larger machine, the 6900, which had 130 hp available and featured a beefed-up front axle and extra PTO clutches, as befitted a tractor that effectively replaced the larger 7600 machine.

1996 saw further new models appear with the 7000 range being updated into the 7010 line-up with the 140 hp 7610, 155 hp 7710 and 175 hp 7810 tractors. These would go on to be classed by many people as the best tractors ever built by John Deere, and the 7810 model in particular is still highly desirable to this day. A new feature on these machines was the option of a new front-axle suspension system called Triple Link Suspension, or TLS, by John Deere. Suspension systems were becoming ever more popular among tractor purchasers and John Deere had a superb system with their version, although the early ones did have problems with wear to some degree. Cab suspension would also arrive later, making for an even smoother ride at speed on the road or while working across ploughed fields.

The 6506 arrived in 1995 as something of a bridge model between the 100 hp four-cylinder 6400 and the 110 hp 6600 six-cylinder machines, the 6506 being a 105 hp tractor but with a six-cylinder motor. This power size had been very popular before the 6000 Series arrived and it seems strange that John Deere abandoned it for so long. Admittedly, average horsepower figures for the best-selling tractors of all makes was on the increase, but the fact that the 6506 was introduced proves there was certainly still a market for this type of machine.

The year 1996 also saw the articulated 70 Series bow out in favour of the all-new 9000 range of four models from the 260 hp 9100 to the 425 hp 9400, all coming with brand-new cabs and PowerTech engines.

By 1997 the 5000 range of 55 to 70 hp tractors had arrived, as had the 3000 Series of 55 to 85 hp tractors, which were actually built by Renault in France but featured Deere engines. Renault produced the same tractors in its own colours as the Ceres range.

The 8100 was the smallest of the 8000 Series tractors. Still using the full-frame design principle, the six-cylinder Deere engines on these tractors were mounted well forward to give an excellent weight distribution front to rear.

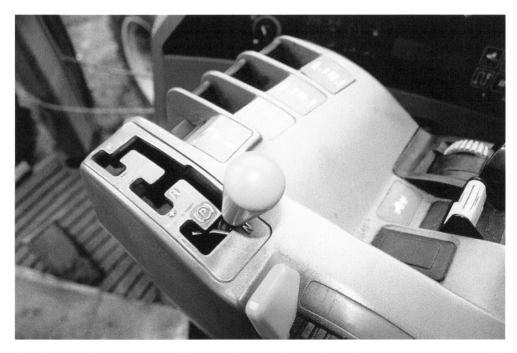

The CommandArm in the cab of the 8100 showing the single small lever used to control the full powershift sixteen-speed transmission.

In 1993 the 6600 and 6800 models joined the line-up, bringing six-cylinder Deere engines into the range of full-frame tractors built at Mannheim. With 120 hp on tap, this 6600 has more than enough power to plant sugar beet in Norfolk with a Stanhay Webb precision drill.

Largest of the 6000 Series was the 6900 with 130 hp on offer. The new 6000 Series styling included a bonnet with removable side panels and hinged top section plus an exhaust stack mounted to the side of the cab pillar.

A John Deere 7810 breaking down ploughed land in East Sussex. This is believed by many people to be the best John Deere tractor ever built! (Photo: Kim Parks)

With more horsepower than before and also brighter new bonnet decals, the 7810 was the largest of three models built in Waterloo, Iowa.

The 6506 was first seen in 1995 and filled a gap in the range with its 105 hp six-cylinder engine, rating above the 6400 but below the 6600.

The right-hand side control console inside the 6506, which was common to all the 6000 Series tractors. The hydraulic raise and lower control, bottom centre, was extremely easy to use and gave fingertip control of the rear linkage.

The biggest John Deere tractor, with the launch of the 9000 Series in 1996, was the 424 hp 9400, an example of which is shown here on a demonstration at a Farm Progress Show in the USA. (Photo: Paul Reeve)

The 5000 Series saw the smaller tractor market addressed, with the 5400 being the largest at 70 hp. Flotation tyres make it look much larger than it actually is.

It was also in 1997 that the 6010 Series replaced the earlier 6000 range. Four- and six-cylinder Deere engines provided the grunt from the 75 hp 6010 up to the 135 hp 6910 and TLS was also now an option. As you can see, most of the range received a boost in power of around 5 hp, but externally they looked very similar to the original range (the main external difference being brighter bonnet decals).

The same year saw the 8000T range arrive, which saw John Deere agricultural crawlers make a return to the marketplace after many decades of absence. Rubber track technology was used to turn the four-model 8000 Series into a crawler version, with the tracks using a large rear wheel and front idler set towards the rear giving a long overhang at the front, where the weight of the forward-mounted engine gave excellent weight distribution for pulling heavy implements. A conventional steering wheel was retained to vary the amount of power to each track and thereby turn the tractor.

The SE economy versions of the 6010 Series were launched in 1999, offering more basic specification and also the option of the standard or low-profile cabs while the 5010 range offered a smaller package of three models.

In 2000 the 8000 Series finally got the 10 Series treatment. The 8010 range featured higher power outputs and new-style bonnet decals.

2001 saw the 6020 Series replace the 6010 models with a new look and many more options on a range that stretched over ten models from the 80 hp 6120 up to the 160 hp 6920S. New-look, curved bonnets were able to be hinged up backwards to give excellent access to the engine for maintenance and were a great improvement over the earlier removable panels. An improved Triple Link Suspension system on the front axle was joined by four-post cab suspension (called Hydraulic Cab Suspension by John Deere). There was also a wide choice of transmission options, including a constantly variable transmission called AutoPower. This was the first time this sort of stepless transmission was offered on a John Deere tractor.

Notable models included the biggest four-cylinder model, the 6420S, whose turbocharged power unit could produce 120 hp plus an extra 5 hp on top at the power take-off. Similarly, the 6920S also had the power boost feature, providing more power at the PTO when needed.

The GreenStar system, which was John Deere's global positioning system, was now fully integrated into the design of the 20 Series tractors, leading to fully automatic working as the tractor drove itself across the field in a dead straight line without driver input.

The new Deere engines in the 6020 Series produced a very notable engine noise while under load or at speed on the road, and there was many a driver who kept the right-hand side window open to enjoy the raucous note of the six-cylinder engine in particular – including me!

Gradually, the 20 number was applied across the ranges, with the 8020 range arriving in 2002 and once again including a new style of bonnet, as well as a new, more powerful 295 hp flagship model called the 8520. A new front axle Independent Link Suspension system as well as fully integrated automatic guidance systems using GPS made these the most sophisticated big Deeres from the USA so far.

1997 saw the 6010 range arrive with the smallest being the 6110; this example of the 75 hp model is shown hedge-cutting in Suffolk.

The 6610 was a 115 hp tractor with its power coming from a Deere six-cylinder motor, which gave more than enough power to use a five-furrow plough in Lincolnshire.

The 6810 was fitted with a heavier 7000 Series type back end to cope with heavier implements and had 125 hp under its bonnet, which was still of the original 6000 Series design.

The Triple Link Suspension system on the front axle of the 6810 used gas accumulators to cushion the ride. It worked extremely well but this early version was prone to wear and was replaced by an updated version in 2001.

The 6910 was a 135 hp tractor but the 6910S gave a little bit more horsepower at the PTO when needed than the standard model. This very tidy example is shown waiting for a load of peas in Kent.

This 8400T, shown working in the USA, was the largest in the new range of four rubber-tracked crawlers based on the 8000 Series wheeled models. (Photo: Paul Reeve)

Part of the SE range of economy tractors, this 6310 is shown at the Cheffins auction in Cambridgeshire. Note the bonnet-mounted exhaust stack used on the SE range.

The 6020 models arrived in 2001 with a totally new look. Here, a Sussex contractor's 6810 and 6920S show the main external differences well. (Photo: Kim Parks)

The 6420S was a powerful little tractor, its four-cylinder engine producing 120 hp and the slightly more 125 hp at the PTO. Although well capable of pulling a large grain trailer, the turbocharger certainly made more noise when that trailer was full!

The 120 hp 6620 offered the same power output as the 6420 but with six-cylinder power and weight. This one was the author's regular steed while carting potatoes from the field to the store in Suffolk.

Inside the cab of the 6620 and the PowrQuad Plus gearbox is controlled by the large lever, with the four powershifts in each range being operated by the 'hare and tortoise' buttons on top. The rest of the tractor's functions are controlled from this console as well, leaving the steering column to be as slim as possible to improve forward visibility.

This 6820 demonstrates the big improvement on routine maintenance on the 6020 range over the older models, thanks to a tilting one-piece bonnet. (Photo: Kim Parks)

The 6920S had more horsepower available at the PTO thanks to its power boost – useful when using a power-sapping rotary cultivator and combination drill to plant wheat after sugar beet in Suffolk.

With 295 hp available, the 8520 is a suitable flagship model and looks impressive bed-forming in Kent. Note the GreenStar satellite receiver mounted in the centre of the cab roof. (Photo: Kim Parks)

The 8420T was the rubber-tracked version of the 8420 and was the third generation of these John Deere crawlers, which were still using a larger rear driving wheel.

The new sloping bonnet of the 7920 made this tractor look very imposing, as did the new cab, which was brought over from the bigger 8020 range and with its larger area of glass gave much better all-round visibility. Wet and sticky conditions mean that this 7920 is struggling to maintain traction while ploughing in Suffolk.

Rubber-tracked versions of these tractors, the 8020T range, were also available. If, however, you required even more power, you could always turn to the articulated four-wheel drive 9020 range, which replaced the earlier monster tractors. There was also a crawler version of the 9020 tractor produced, giving even more choice for those preferring rubber tracks instead of wheels and tyres. Power went up to the mighty heights of 450 hp with these machines.

At the other end of the scale the 5020 range of German-built tractors arrived in 2003 with three models from 72 hp to 88 hp. The same year saw the 7010 range get the 20 treatment as the 7020 Series hit the fields. Built in the States, the 7020 Series featured a brand-new sloping bonnet design and a new cab derived from that used on the bigger 8020 range. These tractors were equipped with the very latest Deere Tier II compliant engines with electronically controlled power boost for PTO and transport applications with a choice of transmissions giving either 40 or 50 km/h top speeds. These included the twenty-speed PowrQuad Plus and AutoQuad Plus, but there was now the option of the AutoPower constantly variable transmission, called the Infinitely Variable Transmission by Deere. The top model, the 7920, was a particularly nice tractor, and once again the engine sounded very distinctive!

The 30 Series arrived in 2005 when the 8020 range became the 8030 line, with the flagship 8530 now producing 330 hp in a rigid-frame design. The 6030, 5030, 9030 and 7030 ranges followed with various upgrades over the earlier models, including new-style bonnets and decals. Inside the cabs, a new light grey replaced the earlier dark brown finish and there was also a new monitoring system controlled by a small screen mounted on the console to the right of the driver.

New 7430 and 7530 models saw the factory in Mannheim producing its most powerful tractors so far. In fact, these were very much hybrid machines, containing components of both the German 6030 and 7030 American-built range. The 7530 proved particularly popular and became the best-selling model of the entire range, followed very closely by the 6930.

The 30 Series saw new styling for the bonnet and new decals, plus an all-new cab interior. This 6430 is turning grass in East Sussex. (Photo: Kim Parks)

This 6630 is working hard with both front- and rear-mounted Baselier potato haulm toppers.

The 6930 saw power increase to 180 hp with power boost for PTO and transport work. (Photo: Kim Parks)

Despite the many options, features and sophistication of the 30 Series John Deere tractors, the cab used on the 6030 and smaller 7030 tractors had now been around since 1992, and indeed the full-frame design first seen then was also still used. New cabs were desperately needed to keep the tractors competitive and this would eventually see the 30 Series being replaced by something that looked rather different.

The biggest of the four-cylinder models in the new 6030 line-up was the 6430, which is shown here hedge-cutting in Sussex. (Photo: Kim Parks)

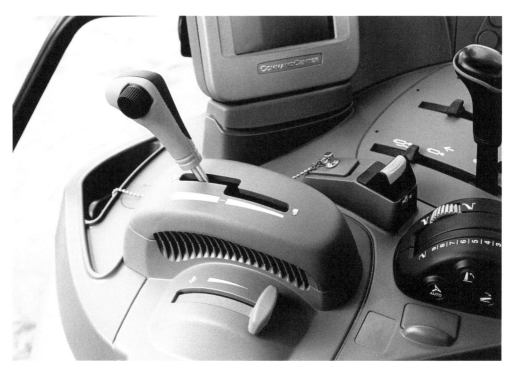

The lighter coloured interior made the 6030 Series much brighter and this single lever was used to control the AutoPower constantly variable transmission, which is seen on a 6930 driven regularly by the author.

The 7530 was a German-built machine using features from both the Mannheim-built 6030 and USA-built 7030 ranges to produce a machine that quickly became one of the best-selling of the entire tractor range.

Inside the cab of the 7530, showing the right-hand side control console, with the lever used to control the semi-powershift transmission taking pride of place.

The CommandCentre computer screen fitted to a 7530 and used for displaying tractor performance as well as various functions.

The 7930 replaced the 7920 but kept the sloping bonnet. A Dowdeswell six-furrow plough is being pulled through hard-baked Suffolk fields by this example.

The 8530 was the new flagship of the American-built 8030 range and still featured a forward-mounted engine.

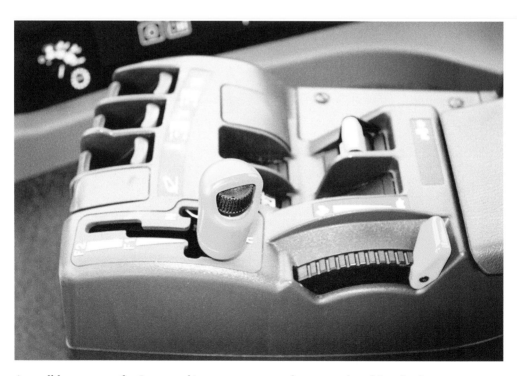

A small lever up on the CommandArm armrest console was used to drive the 8530.

This 9530, shown mole draining in Suffolk, is a very large articulated tractor and has evolved from the 9000 Series design.

The 7530 was an extremely popular tractor and this one is collecting winter barley form a John Deere 9780CTS combine in Suffolk. Deere & Co. was always a full-liner when it came to agricultural machinery and also has a thriving construction equipment line as well.

CHAPTER 7

Thoroughly Modern Deeres

The resultant replacement of the 30 Series was the R Series, which would eventually stretch across the various tractor ranges. New, more aggressive bonnet styling typified these new tractors but it was the brand-new cab that was the big improvement, with a much larger glass area and also more internal space. This was a great improvement over the original TechCentre cab that had survived for so long.

Brand-new engines to meet the latest emissions regulations were included in the new 6R, 7R, 8R and 9R models and a large array of options, including a choice of transmission, were offered. These models also came with a very high level of computer-controlled functions using touchscreen monitors and were also linked to the GreenStar global positioning system, which was even more integrated on these machines than previously.

The current John Deere range is very extensive and includes the three-model 80 to 90 hp 5G range, all coming with four-cylinder engines and no less than four transmission options. Designed primarily with the livestock sector in mind, these tractors nevertheless come with a full range of specifications. Special vineyard and orchard versions are also produced.

The 5M range comprises four models, the smallest of which is the three-cylinder 5075M of 75 hp followed by four-cylinder models from the 85 hp 5085 to 115 hp 5115M, all of which are fitted with a low-profile bonnet for increased vision and also have the option of being factory-fitted front-loader ready.

The 5R range of premium tractors, all using four-cylinder turbocharged Deere engines and stretching from the 93 hp 5090R to the 125 hp 5125R, offer the features usually found in bigger tractors in a more compact package and are once again ideal for loader work.

First seen in 2012, the 6M tractors married the new 6R-type bonnet and engine with the older TechCentre cab design to produce a range of tractors that are a more basic alternative to the 6R models (although the words basic in this context is a rather relative term). A total of eight models make up the range, including both four- and six-cylinder machines from the 110 hp 6110M up to the 195 hp 6195M and with up to five main different gearbox choices, each with many variants within them, including the sixteen- or twenty-four-speed PowrQuad Plus, the twenty-four-speed AutoQuad Plus, the AutoQuad Plus EcoShift, the CommandQuad Plus or the CommandQuad Plus EcoShift. With such versatility, these tractors can be chosen as the ideal tool for any farm.

The 5100R is a 100 hp tractor in a range of what is now considered to be smaller sized tractors.

The M Series of tractors, such as this 6170M, provides an economy version of the 6R tractors and uses the older cab fitted to the earlier 30 Series machines.

The 6MC range of tractors is made up of six four-cylinder tractors from the 95 hp 6095MC to the 115 hp 6115RC, and once again feature a wide range of transmission choices.

The latest 6R models, which form the backbone of the modern John Deere line-up, is split into two groups of either large-frame or small and mid-size-frame machines. The large-frame machines are all six-cylinder tractors with power ratings ranging from the 175 hp 6175R up to the 250 hp 6250R – the latter reaching 300 hp with power boost! The usual range of various types of transmissions can be fitted to these tractors and are joined by the AutoPowr constantly variable transmission and the DirectDrive mechanical equivalent.

The smaller 6R tractors also feature state-of-the-art technology and include models with both four- and six-cylinder Deere engines, stretching from the 110 hp 6110R to the 155 hp 6155R. All have the same wide range of transmission choices and sophistication as the large-frame 6R models.

The big 7R Series tractors are all six-cylinder models stretching from the 210 hp 7210R to the 310 hp 7310R, all of which are available with full electronic systems and a choice of CommandQuad EcoShift with Efficiency Manager, e23 transmission with Efficency Manager or the AutoPowr constantly variable transmission.

The CommandView III cab on the 7R tractors is a huge plus point with plenty of space and excellent visibility all around the tractor.

A six-cylinder Deere engine provides the power for this 6130R, seen planting sugar beet in Suffolk.

The 8R and its 8RT rubber-tracked alternative are the latest version of the original 8000 Series, but have been updated for a modern age. The usual gadgets, including auto guidance and electronic control, are included in the latest up-to-date package. The power ratings have increased as well, with this six-cylinder range starting with the 245 hp 8245R and finishing with the awesome 400 hp 8400R, which takes the rigid tractor over the 400 hp barrier! Only three models are offered as RT rubber-tracked versions: the 8320RT, 8345RT and 8370RT.

You would think this would be the ultimate size of John Deere tractor but you would be wrong in that assessment as the 9R articulated models take the power envelope even further, with the 370 hp 9370R being the smallest and the 620 hp 9620R topping out the six-model range. This range also includes heavy-duty scraper versions of some of these models, with the awesome power of these giant tractors being taken through an eighteen-speed full-powershift transmission. Tracked versions of some of the models are also available, having been fitted with twin sets of rubber tracks. The 9470RT, 9520RT and 9570RT carry on the tradition of these large crawlers, which was first seen on the 9020 Series machines.

New, however, is the 9RX Series, which, although based on the 9R tractors, is fitted with four sets of rubber track units to make a very manoeuvrable crawler, and one that does not disturb the soil when turning on the headland, as a normal twin-track machine tends to do. Four models are offered from the 470 hp 9470RX to the 620 hp 9620RX. With a mixture of high horsepower, crawler traction and enormous pulling power, these tractors could well be thought of as the ultimate John Deere tractor ever; but, as we know, the world of modern tractors never stands still for long.

A 6190R tipping its trailer load of seed for dressing in a Suffolk farmyard. It shows off the new aggressive styling of the R Series tractors very well.

The 6210R has become a very popular model in the range and this one is carting sugar beet with a Bailey trailer. The new cab on these tractors was very expensive to design but well worth it in terms of space and operator comfort. (Photo: Kim Parks)

The USA-built 7230R is a large machine by standard, sitting high and with a comparatively short wheelbase. Dust flies as oilseed rape stubble is given a disc harrowing in Suffolk.

A side view of a 7280R on display at a show in the south of England. (Photo: Kim Parks)

Sharing the same platform, an 8345RT meets an 8320R at a demonstration in Kent.

The 8345RT saw track suspension included on a John Deere crawler for the first time, which improved the ride quality for the operator greatly – especially while at speed on the road. (Photo: Kim Parks)

The new CommandArm control console, including the new screen on the 8345RT.

An 8335R taking a break from work in Kent. The cab was of a new design and gave more room than previously. (Photo: Kim Parks)

The RX machines are the latest to join the John Deere tractor range, which combine the 9R articulated tractor's power and manoeuvrability with four independent rubber track units that prevent the scuffing of the soil when turning on headlands.

This 9570RX is on display at an agricultural show, displaying the huge radiator grille and rear-mounted fuel tanks of these monster machines.

CHAPTER 8

Legacy of the Leaping Deere

Since the takeover of the Waterloo Boy to the latest R Series machines, John Deere tractors have come a long way. During their long history, many of the models produced have gained something of a cult following, especially those such as the 7810, which is seen as the ultimate tractor of all time and today changes hands for a lot of money given their age. The two-cylinder generation is also well remembered thanks to a great number of preserved machines in private collections the world over, and this is helped by the popular Two-Cylinder Club in the USA.

For a long time, John Deere has been a 'full-liner', producing not only agricultural tractors but also ploughs and tillage equipment, planters and harvesters, including combines, forage harvesters and cotton pickers. Other companies have emulated this formula, but few with such success as John Deere, who originated the concept.

This formidable line-up of John Deere tractors at a show in Lincolnshire consists of many different generations of machines, from two-cylinders to 30 Series.

With new machines being produced every year and huge numbers of the older John Deere tractors still out there working for a living on farms all over the world, the John Deere name is sure to remain prominent in agriculture forever. The fact that it has remained a completely independent business during its entire existence is also a lasting testament to the man who started it all that time ago. That is, of course, the man himself – John Deere.

Three two-cylinder tractors, including the Model D that helped establish John Deere as a tractor builder, a Model B row crop model that made the John Deere tractors so successful, and an 830 Diesel, which was one of the last to feature a two-cylinder engine. All are preserved in the UK.

For similar titles please visit:
www.amberly-books.com
The fastest-growing local history, general history, transport and specialist interest history publisher in the UK.